OLYSLAGER AUTO LIBRARY

Fire-fighting Vehicles 1840-1950

compiled by the OLYSLAGER ORGANISATION

research by Denis N. Miller

edited by Bart H. Vanderveen

FREDERICK WARNE & Co Ltd
London and New York

Library of Congress Catalog Card Number 79–186748

ISBN 0 7232 1464 6

Filmset in Great Britain by Keyspools Ltd, Golborne, Lancs.
Printed in Great Britain by BAS Printers Limited, Wallop, Hants

3A

3B

3C

INTRODUCTION

The development of the modern fire appliance is both fascinating and complex. The earliest organized brigade is believed to have been the Corps of Vigiles, operating in Rome before the birth of Christ, but with the fall of the Roman Empire some 500 years later, professional fire protection methods were virtually forgotten until the seventeenth century. It was then that the first manual pumps, or 'engins', were introduced, worked by two or more persons, usually volunteers, water being directed at the blaze by means of a 'squirt' or, as it is now known, a 'monitor'.

By the mid-nineteenth century, prompted, especially in London, by the Great Fire of 1666, several insurance companies had organized their own private brigades which would turn out to premises insured with each particular company. Thus, improved equipment and better organization resulted as each brigade had to show its greater efficiency in order to secure policy-holders from its rivals. Gradually, resources were pooled and the old pedestrian-hauled pumps gave way to horse-drawn units, although still with manual equipment.

The nineteenth century saw the introduction of horse-drawn steam pumping apparatus, not only in Great Britain but also in the Netherlands, Germany and other parts of Europe. The horse-drawn chemical appliance and the first self-propelled steam pumpers were also developed, but the latter were rapidly superseded by the motor-driven appliance which, by 1904, was gaining favour in all fields. Electric traction also enjoyed brief prominence but, like steam, was soon forgotten.

As the 1920's gave way to the Thirties and the Forties, many design improvements were evolved, new hazards discovered and new techniques developed to fight fires. Immediately before, and during the years of World War II, there were many important new developments as a result of exceptional operating requirements and conditions, and these were incorporated in early post-war machines, predecessors of the latest 'limousine' appliances.

With such a complex history and wide diversification of vehicle types, therefore, it has been possible to include only a random selection of typical models within this volume. Nevertheless, we believe that we have captured not only a representative cross-section of the more common models, but also many of the more unusual types which never attained standard production stage.

Piet Olyslager MSIA MSAE KIVI

3A: 1905 Merryweather motor tender of the London Fire Brigade.
3B: 1932 Bedford 2-tonner with Dennis trailer pump.
3C: 1950 Belgian FN Model 62C with Metz turntable ladder.

EARLY DEVELOPMENTS

4A: In 1678, Dunstable in Bedfordshire, purchased a crude 'engin' built by Robert Keeling, a London engineer. Unlike the revolutionary Dutch appliances developed by Jan Van der Heyden, which included hoses amongst their equipment, Keeling's model had to be supplied by bucket chain, making pumping a lengthy and laborious procedure. This machine is now stored at the London Museum.

4B: The horse-drawn steam pumper was invented by John Braithwaite and John Ericsson, partners in a London engineering concern, in 1829. Although three or four similar machines were built, the venture met with determined opposition and eventually failed.

4C: William Rose, Chief Officer of the Manchester Fire Brigade from 1828 until 1846, perfected a particularly unusual type of 'fireman's elevator' (or water-tower) in 1833. It was horse-drawn and consisted of a 2-piece hinged tower with a special 'cage' at its upper end. This could be erected to a height of 40 ft by means of a hand-winching system and water could be directed onto a blaze from the top of the tower. It was not generally adopted.

4D: The 'Winchendon' manual appliance was built by the now world-famous Merryweather concern, of Greenwich, London, circa 1857. It could be adapted for hand or horse draught. This example, now saved for preservation, is fitted out for pedestrian haulage.

4A

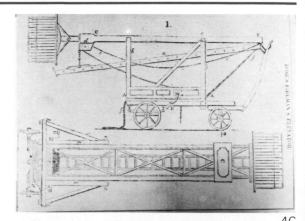

4C

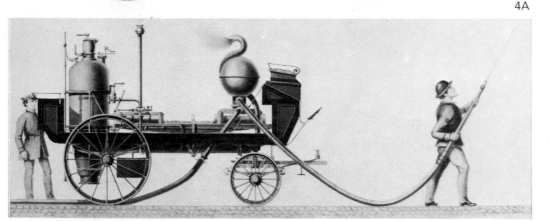

4B

4D

EARLY DEVELOPMENTS

5A: Messrs Hadley, Simpkin & Lott were the predecessors of Merryweather & Sons. One of their many horse-drawn manual appliances was hauled through London to mark the centenary of the great Tooley Street Fire of June 1861. Here, the machine is seen in Tooley Street where pumping has commenced.

5B: The first Merryweather 'Sutherland' steam fire appliance gained First Prize in the National Steam Fire Engine Contest at the Crystal Palace in 1863 and was purchased immediately afterwards by the Admiralty for use in Devonport Dockyard. It could maintain a steady jet of between 160 and 170 ft high. In 1905 it was returned to the manufacturer and, following restoration, was presented to the London Science Museum in 1924, where it now lies.

5C: The police fire brigade was a common sight throughout Britain up to the 1930's. Here, two constables are seen manoeuvring a ladder cart carrying a 3-section extension ladder. These carts were particularly popular in cities and were very similar to the stretcher carts used by the police authorities to carry drunks or dead bodies!

5D: In 1886 a Mr Glenister, Chief Constable and Chief of the Volunteer Fire Brigade at Hastings, Sussex, devised a unique 'first-aid' tricycle carrying, and powered by, two firemen.

5A

5B

5C

5D

6A: The pedestrian-hauled hose reel cart was also very popular at one time. This Merryweather version doubled as a hose-layer, a continuous length being paid out from a reel mounted on the rear axle. The ornate paintwork and other embellishments were unusual for this type of appliance.

6B: As well as the tricycle appliance there were one or two examples of the quadricycle machine. This unique hose-carrier was supplied to Hong Kong by Merryweather & Sons in 1889. Hoses were carried in the central box and it would appear that the bell rang itself with the assistance of vibration!

6C: Many of the larger towns and cities set up special street escape stations in their more congested sectors. At each point an escape ladder was parked and any member of the public assisting a police officer to push the escape to an alarm call would receive a nominal payment, usually one shilling, for his trouble. This photograph shows one of these escapes in Liverpool.

6D: This horse-drawn escape ladder/water-tower shows clearly how the large escape wheels were evolved. The front portion of the escape carrier could be readily detached from the ladder as soon as the appliance arrived at the scene. This is a 1901 Merryweather.

6A

6C

6B

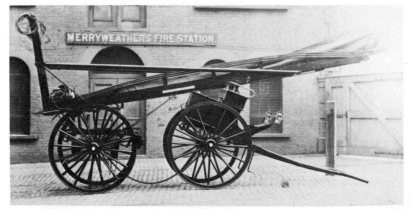

6D

EARLY DEVELOPMENTS

7A: This horse-drawn chemical appliance was originally a manual pump unit supplied in 1888, converted by Merryweather in 1902. The 40 gallons of water in the copper cylinder were expelled under pressure of carbon dioxide gas.

7B: The turntable ladder was first patented in 1888, but did not appear in Britain until 1904, when the Manchester Fire Brigade took delivery of an 82-ft gas-operated unit of Magirus design, manufactured under licence by John Morris & Sons, Salford, Lancashire.

7C: The Manchester Fire Brigade also received one of the first British emergency tenders, a horse-drawn 'light and air' engine by Merryweather. It was ordered by Chief Officer G. W. Parker following an incident at the Shudehill Fish Market in 1900 when almost the entire brigade strength was overcome by fumes. The appliance supplied air to special smoke helmets, and current for handlamps.

7D: Typical of so many British brigades at the turn of the century was that of Swanage, Dorset. The horse-drawn steamer was a new Shand-Mason vertical model, and the wheeled escape ladder was propelled through the streets by hand.

7A

7C

7B

7D

FIRST STEAM FIRE ENGINE CONSTRUCTED IN THE UNITED STATES OF NORTH AMERICA. A.D. 1840.

PAUL R. HODGE·C·E·INVENTOR AND CONSTRUCTOR·NEW YORK·

WEIGHT OF ENGINE 2 TONS, 1 QR, 8 LBS.

DIAMETER OF TWO STEAM CYLINDERS 9 INS X 14 INS STROKE.

DIAMETER OF TWO PUMPS 8½ INS X14 INS STROKE.

QUANTITY OF WATER THROWN OUT OF A 2½ INS NOZZLE 10,824.37 LBS PER MINUTE OR 289.9 TONS PER HOUR TO A HEIGHT OF 166 FEET.

Whilst the self-propelled steam pumper was popular with those brigades that were wary of the internal combustion engine, it was never in very great demand. The world's first is claimed to be that built by P. R. Hodge, of New York, in 1840. It bore a close resemblance to the early railway locomotives of that period, and while pumping was in progress the driving wheels were jacked clear of the ground to act as flywheels.

The first British appliance of this type was built by William Roberts, Managing Foreman of Brown, Lennox & Company, chain, cable and anchor manufacturers, of Millwall, London, in 1862 and was shown at the International Exhibition, Hyde Park, that same year. In practice, it was too heavy, cumbersome and unreliable, and was not generally adopted.

By the 1890's, however, many of Europe's leading appliance manufacturers were experimenting in this direction, and with the repeal of the crippling 'Red Flag' Act in 1896 British companies, such as Merryweather & Sons and Shand, Mason & Company, the latter of Blackfriars Road, London (absorbed by Merryweather in 1922), were able to offer production models, the first of which was a Merryweather 'Gem' for Port Louis, Mauritius, in August 1899.

The spectacle of these machines racing to an alarm call was awe inspiring. With furnace aglow, smoke belching forth and crew clinging like limpets, the appliance would thunder along the streets to the accompaniment of the traditional call, 'HI-YA-HI' from the crewmen, a legacy, it is said, of the days when firemen were all former seamen. The use of bugles, post-horns and even police whistles was not uncommon, but by 1905 the gong or bell, operated by the officer on board, had become more or less universal.

Once the petrol appliance had been developed, the self-propelled steamer stood no chance. Not only did it require readily available supplies of fuel (coal, paraffin or oil), but it had also to be maintained permanently 'in steam' to be of any use on an alarm call. True, some manufacturers marketed special boiler heaters designed to be built into a fire station, but this meant further expense. Thus, the internal combustion engine won the battle.

Despite this, a few steam-driven appliances were delivered to British brigades as late as World War I, and it is even recorded that one or two, along with their horse-drawn predecessors, fought alongside utility machines during World War II.

8A: The works of Paul R. Hodge, designer and builder of the world's first self-propelled steam pumper, were in the now famous Wall Street, centre of American finance. In this early print the appliance is seen in full pumping order, with driving wheels jacked clear of the ground and both cylinders (on either side of the boiler) in operation. The engine weighed just over 2 tons.

9A: William Roberts's 'Princess of Wales', the first British self-propelled steam pumper, was a 3-wheeled vertical-boilered machine weighing 7.75 tons, with a maximum speed of 18 mph. A pulley on the engine shaft could be used for driving fixed machinery as shown.

9B: The most popular Merryweather self-propelled machine was the 'Fire King', of which a large number were shipped to the Colonies. This one was delivered to Singapore during the early 1900's, an unusual feature being the iron wheels to combat damage caused by white ants. Hoses were carried in the storage compartment, which also carried the driver's seat, and water was stored in tanks on either side of the boiler.

9C: Machinefabriek W. A. Hoek (pronounced 'Hook') is now a leading supplier of industrial and medical oxygen gas in the Netherlands. In 1913, however, following many successful years as manufacturers of horse-drawn fire-fighting equipment, they developed two prototype self-propelled pumpers. Both were oil-fired, of 400-gpm capacity and capable of projecting water to a height of over 260 ft.

9A

9B

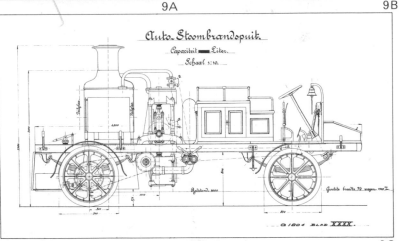

9C

HOSE TENDERS AND OTHER LIGHT 'FIRST-AID' APPLIANCES

The first self-propelled hose tenders, built between 1901 and 1906, were no more than light car chassis or vehicles of similar dimensions, carrying a large hose reel, various branchpipes, nozzles, etc., and occasionally ladders and short lengths of large-bore suction hose. They were developed from the hand- or horse-drawn hose reels and hose carts of the 1850's, themselves introduced for use in areas of constant mains pressure, where hoses could be connected direct to the water supply.

Battery-electric traction was popular for this class of vehicle on the continent of Europe and variations of the hose tender, particularly those of German origin, could cope with twelve or more crewmen. By 1911 electric propulsion had gained some popularity in the United Kingdom also, but only on a relatively small scale, and enthusiasm soon waned.

The 'flying squadron' machine was favoured by American brigades. Again, it was usually based on a touring car chassis but was designed to carry a hand-picked fire crew at high speed in advance of the heavier-equipped appliances. Fire-fighting apparatus was generally restricted to a few hand extinguishers and a box of tools.

10B

10C

10A

10A: Following the success of the Merryweather self-propelled steam pumper delivered to Port Louis, Mauritius, in 1899, the brigade took delivery of one of the first motor hose tenders built by that Company some four years later. Powered by a twin-cylinder engine, it was designed to carry a crew of four or five, 1000 ft of hose and various standpipes, branchpipes, etc.

10B, 10C, 10D: At the Royal Horticultural Society's Show, Holland House, Kensington, in 1906, Merryweather & Sons were awarded the Silver Knightian Medal for this unusual piece of equipment. Designed mainly for country estates, it could be used to harness the power of any motor vehicle should a fire occur. The illustrations show the appliance on its special wheels towed behind a 15-hp touring car (10B); having set up the equipment, the car is reversed up the ramps (10C); full throttle is applied and the car's driving wheels supply the power for the pump which ejects a single jet of water at high pressure (10D). The car was a French CGV, built by Charron, Girardot et Voigt.

10D

HOSE TENDERS AND OTHER LIGHT 'FIRST-AID' APPLIANCES

11A

11B

Early 'first-aid' appliances were similar to the hose tender but carried additional equipment in the form of portable extinguishers, extra lengths of hose and numerous tools. Touring car and light truck chassis were generally popular for such applications.

By 1930 the days of the light hose tender and pumpless 'first-aid' appliance were almost at an end. Few had been built during the previous decade and those that had were fast disappearing in brigade replacement programmes. One combination machine could now take the place of three of these earlier designs.

One quite popular variation of the light 'first-aid' unit was the light pump 'first-aid' appliance, employing a single- or twin-cylinder petrol-powered pump unit which, in some instances, was even mounted on a motorcycle combination.

11C

11A: This 1908 Packard, in the service of the Detroit Fire Department, was typical of the American 'flying squadron' machine. Based on a standard touring car chassis with heavier suspension and special body accommodating fifteen men, equipment included three hand extinguishers and a powerful electric siren.

11B: The model A.3 Albion was a 16-hp machine of conventional layout. The Milton Volunteer Brigade, of Brisbane, Australia, purchased a hose tender based on this model in 1910. In Australia the gong, rather than the bell, was a popular warning device.

11C: Typical of early motorcycle combination appliances was this unit supplied to Parana, South America, in 1913. Conversion work was undertaken by Merryweather & Sons and included a 'Kemik' hand extinguisher, small pump driven from the motorcycle engine, extension ladder and various standpipes, etc.

TD·3866

HOSE TENDERS AND OTHER LIGHT 'FIRST-AID' APPLIANCES

12A: A few years later this slightly more sophisticated motorcycle appliance was produced by Leyland Motors. It was based on a BSA vee-twin machine and featured a special Leyland pump unit with separate engine, ample supplies of suction and delivery hose and a transverse pillion seat for a second man. The pump unit could be removed if necessary.

13A: The Type KZ Renault of 1924 was very popular as a hose and crew tender. This example could carry six men and haul a trailer pump. After a few years it was converted into a light motor pump but a water tank was not included in the specification.

13B: The Model T Ford appeared in many guises during its nineteen-year production run. One example of the 1-ton TT truck, introduced in 1917, was supplied to Bussum Municipality, in the Netherlands, for use as a crew and equipment tender. The trailer pump was a converted horse-drawn appliance.

13C: The Ford Model AA was announced eleven years later. This 12-man hose and ladder truck, with continental style rear hose reels, went into service about that time with Soest Fire Brigade, also in the Netherlands.

13D: Motorcycle combinations were still popular during the Thirties and were now much faster and certainly more reliable than their predecessors. A 1932 Matchless single-cylinder machine was the basis of this Dennis 'first-aid' appliance. It carried a portable pump, suction and delivery hoses and nozzles.

13A

13C

13B

13D

CHEMICAL APPLIANCES

14A

The chemical appliance was introduced towards the end of the nineteenth century to overcome the delay in getting manual or steam pumpers to work. Early types were conversions of the old horse-drawn manuals which, by the turn of the century, had become hopelessly outdated.

The basis of the early chemical machine was usually a large copper cylinder, containing up to 60 gallons of water, normally located in a longitudinal position at the rear. Later combination machines using the chemical unit as a 'first-aid' apparatus, carried this in a transverse position beneath, or to the rear of, the driver's seat.

Generally, on turning a valve, the chemical extinguisher would be brought into play by the reaction of sulphuric acid with a bicarbonate solution. This would generate sufficient carbon dioxide gas to expel the water under pressure.

14B

14C

14A: Powered by a 4-cylinder petrol engine, this early Merryweather appliance featured a copper chemical cylinder located longitudinally at the rear with hydraulically-operated hose reel mounted above. It was claimed that a water jet could be thrown within 5 seconds of arriving at a fire, with all or part of the hose uncoiled, as required.

14B: One of the first carbonic acid appliances was a complicated-looking piece of machinery supplied to an Austrian brigade in 1903. The 123-gallon water tank was located at the rear with two carbonic acid cylinders to one side. A demountable hose reel was carried at the front. Propulsion is believed to have been by 'Knaust-Elektro' system.

14C: This long-wheelbase machine was built by Süddeutschen Automobilfabrik, Gaggenau, a forerunner of Daimler-Benz, in 1907. Classed as a 3-tonner, the 'Grunewald' was offered with engines of 22, 32 or 40 hp, and featured comfortable crew seating, fixed and demountable hose reels.

CHEMICAL APPLIANCES

15A

Unfortunately, this action could not be quelled until the tank was empty, for once the reaction had started it could not, of course, be stopped. Later designs overcame this by having twin tanks, one containing water and the other compressed air. With a tap between the two, the supply could be regulated at will. Alternatively, a single tank of water could be linked with one or more bottles of carbonic acid to produce a similar result. Less common were machines using dry powder-type chemical equipment, designed primarily for combating electrical blazes.

As a 'first-aid' appliance, the chemical machine was most effective. The chemical apparatus, in fact, was still specified by many American brigades as late as 1930, and some are still in use.

15B

15A: In 1907 the City of Ventnor, New Jersey, took delivery of a self-propelled chemical appliance based upon the first commercial model announced by Packard two years earlier. Powered by a 15-hp 2-cylinder engine, providing a maximum speed of 25 mph, it featured coachwork by James Boyd & Bros, of Philadelphia, two 60-gallon chemical extinguishers and various items of hose equipment.

15B: The Radnor City Fire Company, of Wayne, Pennsylvania, included some fifty millionaires amongst its volunteers. A combination chemical and hose car was supplied in 1908 by the Pope Manufacturing Company, Hartford, Connecticut. Two chemical extinguishers were located under the driver's seat and extension ladders hung along the body sides.

16A: The Autocar Equipment Company, of Buffalo, New York (no relation to what is now the Autocar Division of the White Motor Corporation), built this chemical machine for the H. E. Stokes Fire Company No. 3, of Ocean Grove, in 1909. The 40–50 hp chassis carried between twelve and sixteen men, and was equipped with twin chemical extinguishers and 150 ft of hose.

16B: Denver Fire Department purchased a 45-hp chemical and hose car from the Seagrave Company in 1910. It was designed to carry 1500 ft of standard 2½-in fire hose, one 12-ft ladder and a 20-ft extension ladder, plus all the usual equipment. The 40-gallon chemical extinguisher was coupled to 150 ft of hose carried in a special rack at the forward end of the body.

16A

16B

CHEMICAL APPLIANCES

17C

17A

17B

17A: An example of the battery-electric chemical machine is here seen ascending Dartmouth Hill, Greenwich, London, with a full complement of men and equipment. Built by Merryweather & Sons, electrical equipment consisted of a light EPS traction accumulator slung beneath the centre of the vehicle, supplying current to two motors located just ahead of the driving axle. The rear of the body contained a 35-gallon chemical extinguisher and a compressed carbon dioxide gas cylinder, the former coupled to 120 ft of 1-in diameter hose.

17B: The carbonic acid appliance changed little in layout over the years. Compare this 36-bhp Austro-FIAT of 1925 with the 1903 version (Fig. 14B) and it will be seen that basic equipment was very similar. Tank capacity was the same and, once again, two acid cylinders were used.

17C: The Austrian Knaust powder extinguisher of 1935 was also mounted on an Austro-FIAT chassis, a 50-bhp AFN model. A fixed hose reel was mounted on each side and there was suitable locker space for protective clothing and other special equipment.

PUMPERS AND MOTOR PUMPS

The first petrol-driven motor pump, a Merryweather 'Hatfield' reciprocating unit, was built to the requirements of the French Rothschild Estate early in 1904, and within eight months the first combination pump escapes had appeared. Despite the considerable potential of the motor pump as a transporter for the wheeled escape ladder, chemical extinguisher and the like, the 'unadulterated' motor pump unit remained on the scene until after World War II.

There were many variations. Some earlier types were assembled from the rear half of an old horse-drawn steam pumper and front section of a petrol-engined machine, whilst others relied upon battery-electric propulsion. These electric models were sometimes 'married' to an old

18B

18A

18C

18D

18A: The Delahaye-Farcot of 1907 was a best-seller in its field. With a carrying capacity of fiteen men, three large hose reels and pump equipment, examples were sold to many brigades, both in its native France and further afield.

18B: In 1907 the former American motor-racing ace, A. C. Webb, converted a standard Thomas Flyer touring car chassis into a motor pumper. The rotary water pump was located at the rear, driven off the transmission line, and tests witnessed by the Buffalo Fire Department proved most successful. By 1908 the Webb Motor Fire Apparatus Company had assembled many more machines of this type.

18C: The first Dennis motor fire pump appliance was delivered to Bradford, Yorkshire, in 1908. The single-stage centrifugal pump was purchased from Gwynne Cars Limited, of Chiswick, London, and the appliance powered by a White & Poppe engine. This was some years before the introduction of the Dennis turbine pump, developed from a design invented by Signor Tamini, an Italian engineer.

18D: A year later saw the introduction of a new Howe combination pumper and hose wagon in the United States, delivered to the Independent Hose Company, of Frederick, Maryland, displacing an old steam appliance and separate hose wagon. It was capable of travelling at up to 35 mph.

PUMPERS AND MOTOR PUMPS

steam pumper, but where local authorities were more 'go-ahead' they often carried an electric water pump which could be plugged into a nearby mains electricity supply and not only pump water onto the blaze, but also recharge the batteries simultaneously.

The old 'Braidwood' body, where the crew stood along the sides or sat facing outwards, was a horse-drawn development from as far back as 1820. It had been designed by James Braidwood, later Chief Officer of the London Fire Engine Establishment, and examples were still being delivered as late as 1939, notably for wheeled escape ladder duties. Although long lived, it was extremely dangerous and there were frequent accidents involving firemen thrown from machines while answering a call.

In 1931, first Darlington and then Edinburgh, took delivery of the first

19A

19C

19B

19A: Over the years the brigade at Harrow, Middlesex, has owned a succession of Merryweather machines. This example was delivered in 1909 and had an output of 400 gpm. In common with all Merryweather models of the day, chassis and running units were designed and built by the Aster Company, of Wembley, Middlesex.

19B: Bergomi SpA, of Milan, Italy, are now well known for their high-performance airfield fire crash tenders. In 1911, however, this Company supplied an unusual motor fire pump, based on a 35-hp Isotta-Fraschini chassis, to the Turin Fire Brigade. It followed the same general layout as other machines of the day, with rear-mounted pump, crew seats facing outwards and ladders carried aloft.

19C: The Waterous Engine Works Company, of St Paul, Minnesota, USA, were builders of fire appliances and equipment for many years up to 1965, when they were bought out by the American Hoist and Derrick Company. This pumper and hose car was delivered to the New York City Fire Department in 1911.

ever fully enclosed appliances. These featured the van-type body, already popular for emergency tender work. In the old 'Braidwood' style the pump had been mounted at the rear. Now it was located alongside the driver with suction and delivery points on both sides of the vehicle and occasionally at the rear. Pump controls were normally only on the near-side.

Another body design of the period was the 'New World' style, again consisting of an open body but with the crew facing inwards. Access was normally by means of a central entrance at the rear and the pump was positioned behind the driver. The main advantage of this layout was that a greater number of crewmen could be carried than in any other design.

20A

20C

20B

20A: It was at about this time that the first Merryweather turbine machines entered service. The first, which was used only as a demonstrator, featured a German Tetzel turbine, later improved upon by inserting new impellers. The pump could be removed easily and was interchangeable with the already popular 'Hatfield' design.

20B: Early versions of Merryweather's 'Hatfield' pump featured a vertical air vessel, which was easily damaged when loading or removing ladders from the rear. Later versions, such as this 1913 model for Amesbury, Wiltshire, sported a horizontal vessel, although even then a few brigades continued to specify the vertical type. The solid disc wheels fitted here were particularly unusual.

20C: This Daimler-Porsche, built for the Austrian Fire Authorities in 1914, was a 60-hp petrol-electric model featuring midships-mounted pump, twin hose reels behind and a demountable reel at the rear.

Time and time again there had been appeals for increased weather protection for crews and, in 1934, following the new enthusiasm for the fully enclosed van-type body, the first 'limousine' appliances were introduced by the London Fire Brigade. No doors were fitted at first, but later half-doors were included, and by 1947 drop windows were common to all types. In general, the pump was located at the rear.

The bombing of civilian targets during World War II accentuated the disadvantages of the motor pump. It was folly to rely upon a mains water supply under such conditions, so the water tender and trailer pump combination was adopted on a large scale. After the war this was developed into the now familiar pump water tender.

21A

21B

21A: Henry Simonis & Company, of Willesden, London, based their appliances on the Commer chassis. One of the first was supplied to the Base MT Depot, Calais, during World War I, where the Army Service Corps ran the first organized army fire brigade. The 'Barnet' low-loading chassis was powered by a 4-cylinder petrol engine.

21B: The demountable hose reel was a common feature of many European fire appliances except British types. To cater for overseas requirements, therefore, British manufacturers had to be in a position to fit such equipment if specified, an excellent example being this Merryweather for the Cordoba Fire Brigade, in Spain. The side-mounted reels were particularly unusual, especially as these increased the overall vehicle width considerably.

22A: An unusual version of the Ford Model T, and one which was even more unique as a fire appliance, was the British long-wheelbase conversion known as the Baico 'Super Tonner'. This 1921 model, with Dennis No. 1 pump, was even more conspicuous in that it had left-hand drive.

22B: The front-mounted pump of this 1922 British-assembled FIAT F2 1-tonner was driven direct from the road engine crankshaft, a practice developed in Britain and still continued in many parts of the world. Fire-fighting equipment was by Dennis Bros.

22C: The Southampton Fire Brigade's Dennis-equipped Oldsmobile was similar in layout to Romford's FIAT. Again, the front-mounted pump was driven from the road engine crankshaft, but because of the machine's small overall dimensions, lengths of suction hose had to be carried on the ladder.

22D: The Model T Ford, such as this 100-gpm motor pump of 1924, was suitable for all manner of applications, including both rural and industrial fields. Even appliances of this size could supply a good head of water for two hoses.

22A

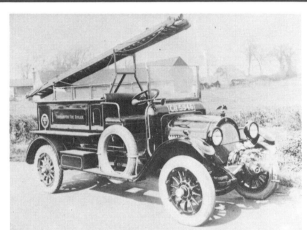

22C

22B

22D

PUMPERS AND MOTOR PUMPS

23A: This smart American LaFrance combination appliance, with motor pump equipment and chemical extinguisher, is still owned by the Renfrew Fire Department, after 44 years service. Nowadays it is retained mainly for promotional activities, fire prevention duties, etc.

23B: The Ahrens-Fox appliance was regarded as 'the Rolls-Royce of the fire service'. Easily recognized by the huge 750-gpm pump ahead of the engine, this combined pumper and hose car left the Company's Cincinnati factory in 1927.

23C: Appliances by Daimler-Benz are particularly popular in Germany today. In 1928 they were equally popular. This machine was supplied to the Stuttgart Fire Brigade at that time. Power was supplied by a 68-bhp engine, which also drove a 440-gpm pump.

23A

23C

23B

24A: A 6-stage 132-gpm pump was located amidships in this 60-bhp Austro-Daimler supplied to the Vienna Fire Brigade in 1930. Hose reels were not carried. Instead, hose was rolled and stowed in special side racks.

24B: A similar hose stowage arrangement was employed in this 45-bhp Gräf & Stift, also of 1930 vintage. The 4-stage front-mounted pump provided 176—220 gpm and a full complement of ladder equipment was carried.

24C: From time to time even the Rolls-Royce car was adapted for fire appliance use. A few Dennis-equipped models went to the Hong Kong Brigade during the Thirties, at the same time as this 'Phantom I', with coachwork by Bonallack & Sons Limited, went to Borough Green & District.

24D: A superb example of the 'New World' body design was a forward-control Dennis FS.6 purchased by the Birmingham Brigade in 1930. It was a particularly compact design, with 900-gpm mid-mounted pump, twin 'first-aid' hose reels and accommodation for a 12-man crew.

24C

24A

24B

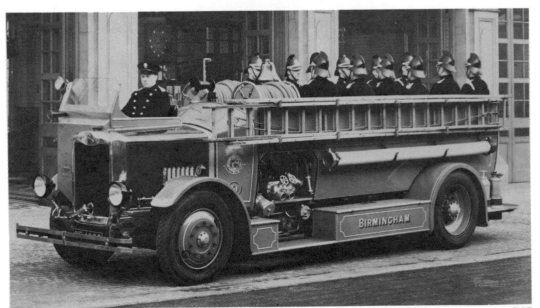

24D

25A

25C

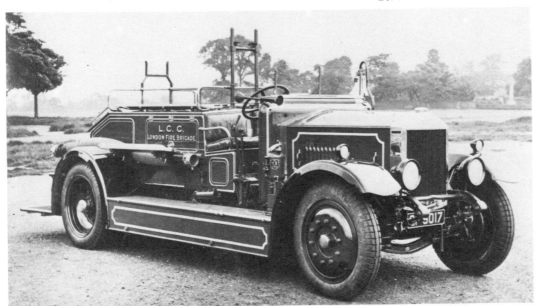

25B

25A: This Mercedes-Benz LS1 appliance had particularly clean uncluttered lines. The major part of the body was taken up with crew accommodation and equipment lockers. At the rear were fixed and demountable hose reels and a 220-gpm pump driven from the engine power take-off.

25B: Between the wars, Merryweather & Sons based their 'standard' appliances on Albion chassis. This London Fire Brigade appliance was typical of many, powered by a 67—70-bhp 4-cylinder engine, and with a 700-gpm rear-mounted pump. The body was of the 'Braidwood' type.

25C: The first ever fully enclosed appliance is believed to have been a Dennis for Darlington in 1931. In addition to the rear-mounted pump and associated equipment, it carried a host of other items for salvage and emergency use, access to crew area being from the rear.

26A: The Ward LaFrance Truck Corporation, of Elmira, New York, has specialized in the manufacture of fire appliances for many years. An example of early Thirties' production was a motor pumper for Woodridge, New York, No. 1 Company. Equipment included 'first-aid' reel and tank, extension ladders, fire hooks and hand extinguishers.

26B: This tandem-drive Tatra Model 30 had a power output of 30 bhp and a 20-hp 77-gpm portable pump. No water was carried but numerous hoses, tools and rescue aids were.

26C: Dennis Bros supplied the first 'limousine' appliances to the London Fire Brigade in 1934, subsequent models being delivered both by Leyland Motors and Merryweather & Sons. All were of the same general pattern, with rear-mounted pump, in this case a 900-gpm variable pressure unit, and hose 'flaked' in continuous lengths in a rear storage compartment.

26D: Many American volunteer brigades continue to maintain vehicles of considerable age. For example, the Middlesex County Division of the Washington Volunteer Brigade is still operating a 1936 Dodge 1½-tonner with coachwork and equipment by the Approved Equipment Company.

26A

26C

26B

26D

27A: A more unusual appliance of the 1930's was the Citroën-Kégresse C6 half-track, of which the Austrian Fire Service had a small fleet. A few of these, fitted with 20-hp 77-gpm portable pumps, operated with 132-gallon water tank trailers. The peculiar headlamp arrangement was specially designed for use with front-mounted snow-plough attachments, etc.

27B: The body of this normal-control Dennis 'Ace' was designed by the Chief Officer of the Sonning, Berkshire, Volunteer Brigade and built by Messrs Markhams of Reading in 1938. Equipment included 1700 ft of hose, a 35-ft telescopic ladder, foam-generating apparatus, three lengths of suction hose, a 40-gallon 'first-aid' system and even a miniature canteen.

27C: Although based on an Albion CX.14 chassis, a common basis for the standard Merryweather appliance between the wars, this handsome appliance supplied to the Glasgow Brigade in 1938 was kitted out by John Kerr & Drysdale Limited, another leading fire equipment manufacturer of the day. The smart saloon body even contained a 'first-aid' hose reel, paid out through the forward windows of the body.

27A

27B

27C

28A: Based on a 1941 American Ford 1½-tonner, the Maxim-equipped 500-gpm pumper and hose car is still in evidence in certain parts of the United States. The 100-bhp V-8 engine drove a Hale pump located amidships.

28B: The East Marion Fire Department, New York State, recently sold its 500-gpm General-Detroit pumper'to a local vehicle enthusiast for $75. It was powered by a 112-bhp Hercules engine and equipped with a 2-stage centrifugal-type pump by the Northern Pump Co., of Minneapolis, and a 500-gallon 'first-aid' tank.

28C: After the war, large quantities of German and Allied military vehicles became available throughout Europe at very reasonable prices. Fire authorities were quick to realize the potential of such equipment, particularly those of all-wheel drive configuration. Here, a Canadian Ford F15A 4 × 4 model has been converted into a motor pump unit of 275-gpm capacity for service in Austria.

28D: A much modified ex-WD Canadian Chevrolet 4 × 4 heavy utility truck was the basis of one of the first high-pressure gear pump appliances designed and built by Van Bergen, of Heiligerlee, Netherlands, in 1951.

28C

28A

28B

28D

PUMPERS AND MOTOR PUMPS

29A: Most American pumpers featured mid-mounted pumps, and a large number of machines, even today, were of entirely open design. This Ward La France, delivered soon after World War II, typifies appliances of this period. Equipment was basically the same as that fitted to the earlier model (Fig. 26A).

29B: The wartime all-wheel drive Opel 'Blitz' 3,6–6700A was ideal for fire-fighting use. Powered by a 75-bhp engine, this specimen for the City of Vienna carried a 330-gpm front-mounted pump, 176-gpm portable pump and the usual fully enclosed body of standard continental layout.

29C: A similar layout was used for the larger Mercedes-Benz/Metz LF25 appliance of 1949. Fire-fighting apparatus included a demountable hose reel, fire hooks, ladders and smoke-extraction equipment.

29A

29B

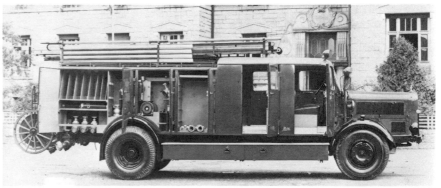

29C

TRAILER PUMPS

It is difficult to ascertain exactly when trailer pumps were introduced. As early as 1888 the German Daimler concern had built a motor fire pump for manual propulsion and sold one to Cannstatt, Germany, in 1892. In 1895 the first motor pump for a British customer went to the Hon. Evelyn Ellis for use on his country estate at Datchet, Buckinghamshire. This was also a Daimler and was shown at the Tunbridge Wells Exhibition and Demonstration of Motor Vehicles on 15 October 1895.

Many brigades, in the British Isles and abroad, subsequently converted their old horse-drawn steam pumpers into trailer units, mainly because early motor-driven machines rarely carried their own pumps. This practice increased considerably during World War I as horses were requisitioned by the military authorities and could not, therefore, haul the old pumpers.

The trailer pump, as we know it, first appeared during the 1920's. Initially it was popular with only a few brigades, principally in rural areas where,

30A

30C

30B

30D

30A: At the Tunbridge Wells Exhibition and Demonstration of Motor Vehicles in 1895, a horizontal single-throw belt-driven pump by the German Daimler concern gave a good demonstration of its capabilities in the hands of the local volunteer fire brigade. Its owner, the Hon. Evelyn Ellis, the bearded gentleman on the right of the machine, is generally believed to have been the first man to bring a car into England and drive it, also in 1895.

30B: Quite a number of old horse-drawn pumpers, particularly in the United States, were adapted for towing behind self-propelled machines. Here, an early American LaFrance steamer is hauled by a 1910 Seagrave chemical appliance.

30C: Amongst the exhibits at the Royal Lancashire Agricultural Show, Burnley, in 1926, was a Leyland trailer pump unit weighing approximately 2 cwt. Not only could it be towed behind a motor vehicle but it could also be carried with ease by four men. The pump shown is believed to be the prototype.

30D: By 1930 an improved design of Merryweather 'Hatfield' trailer pump was available—a 275-gpm model powered by a 40-bhp 4-cylinder petrol engine. The first of these went to Kendal, Westmorland, another customer being Ledbury, Gloucestershire. The Ledbury machine was used in conjunction with a 2-ton Bedford fire tender shown here soon after delivery.

31A

31B

for cheapness and manoeuvrability, and because it could be coupled to virtually any motor vehicle, it was of far greater use than the self-propelled motor pump.

It wasn't until World War II that the trailer pump really came into its own. British self-propelled utility machines delivered up to and during the early war years did not have built-in pumps. Instead, trailer pumps, outnumbering the utility machine by nearly 3 to 1, were manufactured, the idea being that these could be deployed where necessary and were not restricted by heaps of rubble or gaping craters resulting from earlier enemy action.

There were both light and large trailer pumps, the former with a capacity of approximately 100–140 gpm and the latter of between 430 and 500 gpm. They were powered by anything from motorcycle to heavy truck engines. With the cessation of hostilities, and the increasing use of lightweight metals, the trailer unit was largely superseded by the portable pump carried by the self-propelled appliance. This had been perfected in Europe, and in Germany in particular, during the Thirties.

31C

31A: Seven years later Dennis Bros developed a new type of trailer pump to Home Office requirements. Powered by a 70-bhp petrol engine, coupled direct to a No. 2 Dennis pump, it had a capacity of between 350 and 500 gpm. The general layout set a precedent for all subsequent British trailer pumps, incorporating stowage areas for lengths of ordinary hose as well as the suction type.

31B: During World War II, many fire apparatus manufacturers produced light trailer pump units for use with private cars and other light vehicles. In January 1941 Dennis Bros delivered a number of 80–100-gpm trailer pumps powered by JAP single-cylinder 3½-hp water-cooled engines, ideal for ARP (Air Raid Precaution) use.

31C: The US Marine Corps received pumps of more unorthodox pattern that year. These carried twin hose reels at the rear with hose ready coupled for faster laying operations. The unusual wings were of the same pattern fitted to certain of the lighter Dodge commercial models of that time.

PUMP WATER TENDERS

As more efficient replacements were discovered, so the trailer pump vanished from the scene, only a handful remaining in operation by 1950, and these mainly as auxiliary stand-by units.

One such replacement was the pump water tender, a development of the utility water tender and trailer pump combination described later. The PWT (as it is often referred to) carries its own supply of between four and six hundred gallons of water, plus a built-in pump. Thus, it can operate independent of any mains source and is therefore ideal for remote country areas where water is in short supply. This type of appliance, however, has proved so versatile that few brigades, at least in the United Kingdom, do not include the PWT on their strength.

33A

33C

33B

32A: This Merryweather appliance, supplied to South America in 1921, was designed to act as a fire-fighter where water supplies were often non-existent or for use as a street-watering vehicle. The machine was basically a motor pump unit with demountable hose reel at the rear and water tank occupying much of the body area.

33A: This Austrian-operated British Fordson WOT6 4 × 4 machine carried a 440-gallon tank and 330-gpm front-mounted pump.

33B: The single-drive version of the Canadian Ford F60 series was the F602L. This Austrian appliance featured a double cab, 330-gallon capacity tank and 176-gpm pump at the rear.

33C: The 4 × 4 version is here seen with cab in more or less original form, crew area being added on behind and the 440-gallon tank to the rear of that.

C

34A

34C

34B

34D

34A: The lighter wartime Canadian Fords also found good use in the Austrian Fire Brigade. The 1940 F15 4×2 model seen here had a 220-gallon tank and 275-gpm pump, the latter mounted on a special extension frame and driven from the road engine crankshaft.

34B: Dodge D 60 L models were particularly popular for this type of work as shown by this 330-gallon model with 330-gpm pump. This example was unusual in that it had panelled-in body sides for locker space.

34C: The Baden, Austria, No. 2 Company operated this right-hand drive Bedford 'O'-Series PWT with 330-gallon tank, TS8 176-gpm pump, extension ladder and crew cab.

34D: By 1946 several European brigades had the ex-military German Opel 'Blitz' 3,6–6700A four-wheel drive model on their fleet strength. Tank capacity was 330 gallons and pump output 176 gpm.

PUMP WATER TENDERS

The earliest pump water tenders delivered in any great numbers came after the war and were based upon existing utility water tenders or, in many parts of Europe, on reconditioned ex-military chassis. Many were completely re-bodied, often by the brigades themselves, incorporating crew-cabs, numerous lockers, one or two 'first-aid' hose reels driven from the machine's pump at idling speed, and even 2-way radio developed from the wartime VHF system.

Whilst the immediate post-war British PWT normally had a well-panelled fully enclosed body, its continental equivalent; which was often based on an ex-military chassis, frequently carried an ordinary water tank at the rear with virtually no panelling whatsoever, and it wasn't until the very late 1940's that the fully enclosed PWT made its appearance in continental Europe on any large scale.

35A

35B

35C

35A: This combined PWT and street-watering vehicle was also based on the wartime 4 × 4 Opel 'Blitz' chassis but featured a standard cab layout, thus providing additional room for a 440-gallon tank.

35B: A dual-purpose rarity supplied to Turkey in 1949 also served as a street-washer or pump water tender. Body design followed the 'New World' style closely with the tank, of approximately 800 gallons capacity, located within. Basis of the machine was a Dennis 'Pax' normal-control municipal model.

35C: The underfloor-engined 2-stroke Commer truck, unveiled in 1948, was frequently adapted for fire-fighting purposes and eventually resulted in a special fire appliance chassis, the 'A'-Type. A design by James Whitson & Company, of West Drayton, Middlesex, based on the original goods chassis featured a 400-gallon tank, 50-gpm power take-off driven integral pump and light 250-gpm Coventry Climax portable pump.

ESCAPE VANS

The origin of the term 'escape van' seems somewhat obscure, particularly as the first self-propelled wheeled escape units were not vans in any sense of the word. Another dictionary definition for 'van', however, is 'front of an army or fleet when advancing or in battle array'. It is quite likely that this is how the term originated because, not only does it tie in with the fact that the escape vehicle has always been given priority over other appliances when lives have been at stake, but also continues the old naval traditions and terminology which remain the backbone of the fire service.

Early British wheeled escapes, by J. Gregory in 1819 and Abraham Wivell in 1829, were crude, to say the least, and neither was successful. In 1836, however, Abraham Wivell introduced a new design, incorporating the now familiar wheels to aid manoeuvrability, and this he patented. The Royal Society for the Protection of Life from Fire was quick to realize the potential of his invention and adopted it as standard equipment, developing the idea still further.

36B

36A

36C

36A: The world's first motor-driven escape was designed by Supt. S. M. Eddington of the Tottenham Brigade and assembled by Merryweather & Sons. Equipment included a 60-gallon chemical extinguisher, fixed hose reel and 50-ft escape. The Harringay Fire Station, to which it was allocated, was the first in the United Kingdom designed specifically for motorized appliances.

36B: The Bombay Fire Brigade took delivery of a Merryweather escape van in 1904. The wheeled escape's frame was specially sprung and the whole loaded and unloaded by using a block and tackle. This, in itself, was unusual as the normal method of shipping the escape was, and still is, by a balance pivot system.

36C: This 1905 Merryweather combined escape van and chemical extinguisher was based on a similar chassis to the 1904 model but carried the more common balanced escape ladder, with twin chemical extinguishers and 'first-aid' hose reel beneath.

36D: This London Fire Brigade combined chemical extinguisher and escape van was based on the same Merryweather model as the appliance illustrated on page 3. The barbed hook of the Pompier ladder can be seen above and to the left of the escape wheels.

36D

37A: The steam-propelled escape unit was quite a rarity. Merryweather & Sons was one manufacturer who patented such an appliance but few, if any, were built. The design included a vertical boiler as near as possible to the front of the machine to alleviate fouling of the escape ladder.

37B: Cedes was the name given to battery-electric vehicles built by Austro-Daimler, one of the ancestors of the Steyr-Daimler-Puch concern. The London Fire Brigade took delivery of a few examples in the years leading up to World War I. All are believed to have been escape vans with chemical extinguisher 'first-aid' apparatus.

37C: In 1932 the Congleton Fire Brigade had a special escape van built on a Thornycroft BC chassis. The body, by the Lawton Motor Body Building Company, of Stoke-on-Trent, was of the 'New World' type with seating for 14 men, rare for this type of appliance. A 60-ft wheeled escape and 300-gpm Dennis trailer pump were also carried, loading being undertaken from the rear.

An improved design of wheeled escape ladder was announced by James Shand, of Shand, Mason & Company, in 1880, based on Wivell's design, but with a pair of sheer-legs which could be swung out from the ladder to convert it into a free-standing water-tower. This dual-purpose role became unnecessary when, during the Thirties, the large-scale adoption of the German-designed turntable ladder displaced the need for such an appliance and it reverted to its original purpose, that of purely an escape vehicle.

The first horse-draught escape carrier appeared in 1890, and by June 1903 the first motor-driven escape had entered service, at Tottenham, North London. The battery-electric appliance was particularly useful as an escape carrier as no power was necessary to operate the escape equipment.

Between 1932 and the outbreak of war, many brigades converted their old escape vans into combination machines, notably pump escapes, a far cheaper method than purchasing a new motor pump and replacing worn out escape vans. Thus, by the late Thirties, new escape vans were non-existent and many of the old ones had found a new lease of life or vanished completely.

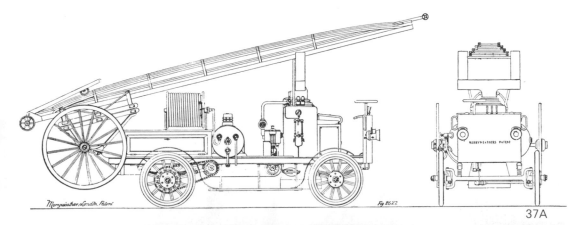

37A

37B

37C

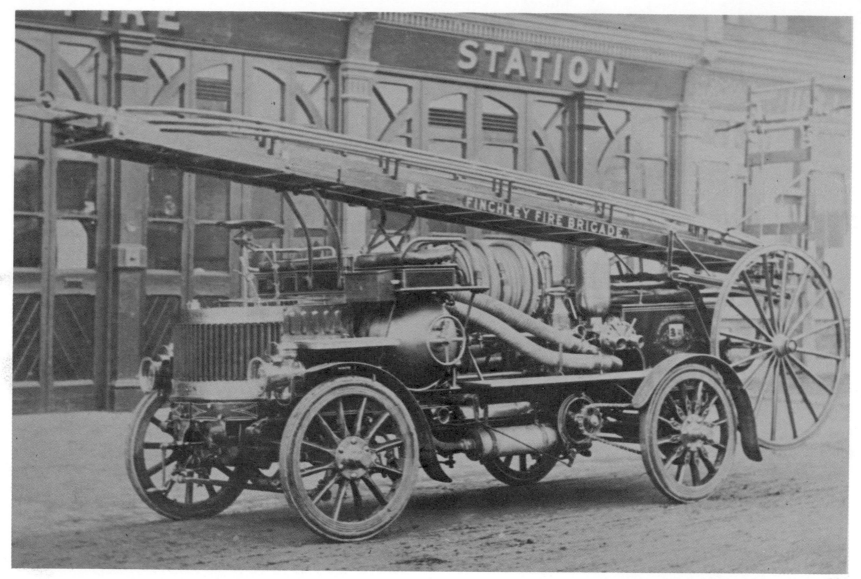

PUMP ESCAPES AND DUAL-PURPOSE MACHINES

39A

Prompted by Tottenham's enthusiasm for the self-propelled escape unit, the nearby Finchley Brigade, under Chief Officer Sly, took delivery of the world's first motor pump escape, also a Merryweather, in November 1904. The original 30-hp 4-cylinder engine was later replaced by a more powerful unit and the appliance, minus escape ladder, now stands in the London Science Museum.

In a number of cases, brigades returned their old motor pumps to the manufacturers for conversion into pump escape appliances and a few even converted these in their own workshops. Since then design trends have followed closely along the lines of the motor pump. There were no pump escapes of the 'New World' type however, mainly because of crew accommodation problems, and the van-type body was similarly unsuitable. The location of pump controls and outlets at the rear was also discontinued because, unless the escape ladder was first removed, it was extremely difficult to reach these controls.

After the war another version of the pump escape, known as the dual-purpose appliance, made its début. Basically, this was a PWT adapted for escape ladder work, and designed principally for small country town operation. Although heavier and not so fast as the two separate machines the advantages, both economically and operationally, of running a single machine were numerous.

39B

39C

38A: The world's first motor pump escape was a triple-combination Merryweather, comprising a 60-ft escape ladder, 60-gallon chemical extinguisher with 180 ft of hose, and a triple-barrel 500-gpm 'Hatfield' pump. Costing nearly £1000 in 1904, it was withdrawn from service in 1928.

39A: By 1906 the same Merryweather chassis (built by the Aster Company) was available for a variety of applications. This example was supplied to the Singapore Fire Brigade and featured a 50-ft wheeled escape and rear-mounted pump unit.

39B: Early Dennis turbine pumpers were especially popular as pump escape appliances. This model was supplied soon after World War I to the Hull Fire Brigade which, in those days, was still under the control of the local constabulary.

39C: Belsize Motors Limited, of Manchester, built one of the most powerful British fire appliances of the day after World War I. Powered by a huge 8-litre petrol engine, this example, with wheeled escape and fire apparatus by John Morris & Sons, was built in 1920.

PUMP ESCAPES AND DUAL-PURPOSE MACHINES

40A: The Dennis turbine pump was developed from a design invented by Signor Tamini, an Italian engineer, and fitted to all subsequent Dennis pump appliances. The local Guildford Fire Brigade took delivery of one example in December 1928. Powered by a 60-bhp engine, the wheeled escape/water-tower was a John Morris 'Ajax' type.

41A: This smart Albion/Merryweather was one of the last of this type to be supplied with solid tyres, going to Bromley, Kent, at the beginning of the 1930's. This particular model was more popular as a motor pump unit, with 'Hatfield' pump equipment, than as a pump escape appliance.

41B: The Hendon Fire Brigade was a regular buyer of Leyland appliances. This turbine pump escape/water-tower unit was delivered in 1930 and featured a rear-mounted pump, 'Braidwood' body and 'first-aid' hose reel.

41C: A 1932 Leyland FT.3A pump escape appliance was completely restored for a starring role in the Associated British Picture Corporation's screen comedy 'Go To Blazes'. Some familiar faces are seen here (l. to r.) in the forms of Robert Morley, Dennis Price, Dave King, Daniel Massey and Norman Rossington. The main disadvantage of a pump escape outfit of this type (with the pump at the rear) is clearly illustrated here, where the wheeled escape unit has to be removed before suction hoses can be unloaded.

41A

41B

41C

PUMP ESCAPES AND DUAL-PURPOSE MACHINES

42A: In 1935 the Watford Fire Brigade placed an order for, and received, its second Scammell appliance, the idea being to keep employees at the local factory sufficiently busy during the 'slump' period. It was appropriately named 'Silver Jubilee' to commemorate the 25 year reign of George V and Mary, and had the usual 'Braidwood' style body mounted on a low-loading chassis. Following periods of employment with Metro-Goldwyn-Mayer and a Middlesex heavy haulage contractor, it was acquired for preservation in 1970.

43A: Seen here before the escape ladder was fitted is a Dennis 'Light 6' limousine appliance of 1937. This machine was of particularly striking design, with pump located amidships and duplicate controls on both sides. Other equipment consisted of a 40-gallon 'first-aid' tank, with hose reel behind the crew compartment, and a flatrailed area at the rear for soiled hose, jump sheets, etc.

43B: Leyland Motors produced large quantities of fire-fighting vehicles during the Thirties and Forties. The Royal Army Service Corps received several of the FKT type, with 500-gpm turbine pumps, 50- and 60-ft escape ladders and numerous extension and Pompier ladders.

43C: After the war many British utility appliances were adapted for peacetime operation. Some, such as this V-8 Fordson escape carrier based at the old Park Royal Fire Station in North-West London, required little or no modification.

43B

43A

43C

PUMP ESCAPES AND DUAL-PURPOSE MACHINES

45A

45B

44A: During the winter of 1947/8 the National Fire Service allocated a new Leyland appliance to its station at Wealdstone, Middlesex. Not only did this machine carry a wheeled escape ladder and mid-mounted pump, but also 300 gallons of water for use in outlying districts. This was one of the very first new dual-purpose appliances in the United Kingdom.

45A: The 162-in wheelbase Dennis F.7 dual-purpose appliance, powered by a Rolls-Royce B.80 Mk. X petrol engine, made its début in 1949, the first going to East Ham, London. Performance figures were truly staggering, with an acceleration time of 0—60 mph in just 45 seconds! Although a number of brigades took delivery of this model, it was not entirely successful because of its long wheelbase. Within a year it had given way to the now familiar F.12 model.

45B: The F.12 was a 150-in wheelbase version of the F.7. Mechanically it was identical to the F.7, with a 900—gpm No. 3 Dennis pump located amidships, but with revised frontal appearance. Production lasted for nine years, later options including a rear-mounted No. 3 pump of 1000-gpm capacity. This version, however, saw more use as a major pump than as a pump escape.

46A

46B

46C

As buildings increased in height, so the need for higher fire escapes grew also. In 1888 a special metal aerial ladder, roughly the American equivalent to the European turntable ladder, was sold by its manufacturer, the E. B. Preston Company, of Chicago, to the city of St. Paul, Minnesota. A short time afterwards, Seattle, Washington, also took delivery but here the idea died, not to be revived until 1936 when Peter Pirsch & Sons, of Kenosha, Wisconsin, announced the first all-metal hydraulically powered aerial ladder.

During this vacant period, the hook and ladder truck coped successfully with all escape duties throughout the United States and Canada. Both this and the later aerial ladders were mounted on rigid or articulated chassis, the aerial ladder turntable being located at front or rear of the machine. Occasionally, combination appliances consisting of aerial ladder, hook and ladder equipment, jump sheets and even chemical extinguishers for

46D

46A: The first British turntable ladder appliance (TL) to use power from the road engine to elevate the ladder was supplied to Shanghai in 1908. This was a Merryweather design and was powered by a 40-hp 4-cylinder petrol engine. Elevation and extension movements were governed by two levers at the rear of the machine and four stabilizing jacks were fitted behind and ahead of the rear axle.

46B: The German Daimler concern based some of their early turntable ladders on the battery-electric Lohner-Porsche chassis. The 84-cell batteries of this 1909 example provided 146 amp/hr capacity and the ladder was a 4-stage unit elevated by the action of compressed carbon dioxide gas.

46C: This 36-hp Benz-Gaggenau was delivered in 1912, ladder equipment being fitted by the German Metz concern. As well as the demountable hose reel located at the rear, this appliance featured a 176-gpm mid-mounted pump unit.

46D: On 16 April 1913 the city of Aberdeen, Washington State, took delivery of a new Pirsch 55-ft city service ladder truck. As well as the numerous ladders and other rescue apparatus carried, the specification called for a small chemical extinguisher and lengths of 'first-aid' hose stored in racks behind the crew. The price was $6070.

AERIAL LADDERS, HOOK AND LADDER TRUCKS AND TURNTABLE LADDERS

47A

47B

'first-aid' use were delivered to special order.

The world's first turntable ladder is believed to have been a 90-ft German Magirus built in 1892 and mounted on a 4-wheeled horse-drawn base. By 1901 elevation was obtained by using compressed carbon dioxide gas, and in 1906 Magirus introduced the first turntable appliance to feature elevation movements powered by the road engine. By 1932 all-steel ladders, providing far greater heights than ever before, began to appear and hydraulic ladder operation was perfected soon after World War II.

The majority of these turntable ladders could, like the old wheeled escape ladders, be adapted for use as static water-towers, water being pumped to the 'monitor' at the head of the ladder by a separate motor pump appliance. Self-contained turntable ladder/motor pump units soon followed, the majority employing the turbine-type pump unit. A special water-tower appliance was developed for this application in the United States and Canada. This was normally a rigid tower which could be raised into a vertical plane by means of a system of hand winches. For many years it was the only method of attacking serious outbreaks from above.

47A: The American LaFrance Fire Engine Company built vast numbers of 'rigid' and 'articulated' hook and ladder trucks up to the late 1930's. One of these, a Type 14 'rigid' model powered by a 4-cylinder engine, was supplied to Hightstown, New Jersey, in 1915, and remained in service until quite recently.
47B: This 1924 Merryweather for the Corporation of Rangoon incorporated yet another untried system of ladder elevation. A 3-throw pump mounted at the rear of the machine was driven by a worm reduction gear from the main road engine. This pump delivered oil at a working pressure of approximately 450 lb/sq. in, thus providing sufficient pressure to raise and extend the 93-ft ladder.

48A: The world's first all-powered aerial ladder truck was manufactured by Peter Pirsch in 1931 and subsequently delivered to Spokane, Washington State. This was a semi-trailer model incorporating steerable rear wheels with rear steersman's seat located above. The aerial ladder turntable was mounted exactly over the coupling and beneath the ladder were numerous other ladders and rescue aids, each for a different application.

48B: Edinburgh Fire Brigade purchased a new Albion/Merryweather TL in 1929, fitted with pneumatic tyres all round. The 3-stage ladder was constructed of wood and steel, with elevating movements powered by a separate engine located at the ladder's base. The appliance was easily adapted for use as a water-tower if required.

48C: This Metz-equipped Mercedes-Benz of 1933 was powered by a 100-bhp engine and weighed some 5 tons. The general layout seen here, with open cab, clear body area and adequate locker space, was popular throughout the Thirties and Forties, especially in the United Kingdom. Ladder movements were powered by the road engine and by this time the all-steel turntable ladder was coming into its own.

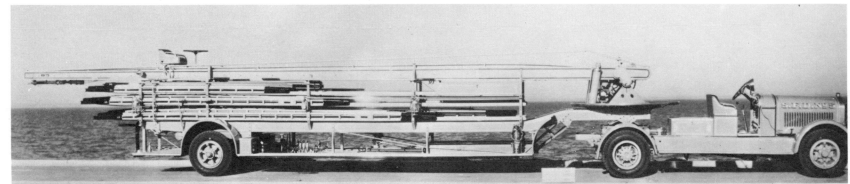

48A

48B

48C

49A: This impressive machine was the first all-British 100-ft motor turntable ladder with steel sections, supplied to Hong Kong in 1933. Raising and elevating movements of this Merryweather machine were controlled by foot accelerator and the whole could be raised and extended inside 25 seconds. Power was supplied by a 170-bhp 6-cylinder engine.

49A

49B

49B: German firms, such as Magirus and Metz, are world famous for their fire-fighting apparatus and for turntable appliances in particular. Photographed in Regensburg, Southern Germany, in 1968, this splendid Magirus is typical of German design and engineering skill of the mid-Thirties. Using a low-loading chassis such as this resulted in a very low centre of gravity and, consequently, an extremely stable machine even in the strongest of winds.

50A: The impressive Pirsch aerial ladder truck delivered to Melrose, Massachusetts, in 1936, was claimed to be the first 100-ft aerial ladder in the United States. Although similar to the Pirsch supplied to Spokane five years previously (Fig. 48A), the new model contained many detail improvements, including all-steel construction, greater horsepower and a generally improved trailer layout.

50B: A 1937 delivery to the Wimbledon Fire Brigade was this 75-ft TL from Merryweather & Sons. Incorporating many of the features found in the 1933 Hong Kong machine (Fig. 49A), a further improvement was

ladder elevation by a single-tension screw mechanism. The comparatively short wheelbase of this example was unusual for this class of appliance in the United Kingdom.

50C: Similar to the Wimbledon model, but with forward-control, was this 75-ft machine supplied by Merryweather in the same year. Power, both for raising and elevating the ladder and for operating the 275-gpm 'Hatfield' pump, was provided by a 100-bhp engine located between driver and mate.

50A

50B

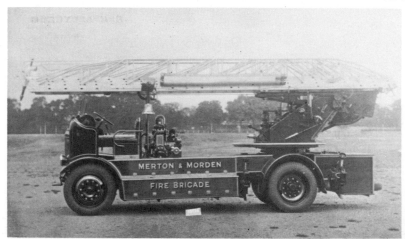

50C

AERIAL LADDERS, HOOK AND LADDER TRUCKS AND TURNTABLE LADDERS

51A: The 1939 American LaFrance illustrated here was supplied to the Chicago Fire Department and, like the earlier Pirsch models, carried a host of other rescue aids beneath the aerial ladder. The closed cab was particularly unusual for this period, but presumably specified to protect the crew from the severe winter conditions prevalent in that city.

51B: When war seemed imminent, brigades throughout Europe began re-equipping their fleets with brand new equipment. At first, these new appliances were more or less the same as their immediate ancestors, but once the war had started all manner of machines were adapted for fire-fighting applications. One of the former was this Dennis TL which, although similar to other Dennis models of this class produced during the late Thirties, included one or two utility features common at this time. Amongst these were the pump located behind the crew area and a drab grey colour scheme. Ladder equipment was by Merryweather & Sons.

51A

51B

51C

51C: The dangers of overloading or using insufficiently stabilized turntable ladders has always been known. Accidents, however, can still happen. This Preston-registered Leyland/Metz 100-ft unit came to grief with near fatal results during the 1940's. Like the Dennis (Fig. 51B), it was one of a number of appliances supplied on outbreak of war which later passed to the newly formed National Fire Service in 1941.

52A: The highest recorded turntable appliances could be elevated to a maximum of 160 ft, 10 ft higher than the previous highest supplied to Hull in 1935. The 160-ft models were based on open-cabbed Leyland units with ladders and other equipment by Metz. Special features included telephonic communication between ground crew and fireman, and 'elephant's foot' stabilizers at the rear.

53A: Once war had been declared, the use of new German fire-fighting apparatus in Great Britain became out of the question. Merryweather & Sons had been producing such apparatus for some years so theirs was an obvious choice for many brigades. Again, the Leyland chassis was especially popular and now incorporated a centrally mounted pump supplying water to the 'monitor' at the head of the ladder. The main advantage of the Merryweather design was that the operator travelled round with the base of the ladder and was always in a convenient position for operating the controls.

53B: The Metz turntable ladder is popular throughout Europe. In Czechoslovakia, for example, this Praga RN, built at the very end of the war, carried a 4-section 100-ft hand-operated unit with no 'monitor' system. If hose was required to tackle a large blaze from above, it would be run up the ladder from the large hose reel at the rear.

53A

53B

FOAM TENDERS

Oil fires are amongst the most hazardous a fireman can be expected to tackle. Water cannot be used as this merely spreads the flames. The only way to extinguish such a blaze is to suffocate it by spreading a blanket of foam over the affected area. Thus, in districts where the dangers of oil are particularly acute, near oil refineries or airfields for example, special equipment is called for in the form of a foam tender.

The earliest foam tenders appeared in 1920 when the aeroplane was beginning to make itself felt in many parts of the world. These tenders were based on standard commercial truck chassis and carried anything up to 500 gallons of special foam compound which, when mixed with water via the various branchpipes carried, produced many thousands of gallons of foam.

54A

54A: Typical of many early foam tenders was this Merryweather design based on a 1928 Albion commercial chassis. Water was carried in a special tank located behind the crew and pumped into a mixing chamber towards the rear. The foam compound, which was carried in cans along both sides of the appliance, was poured into this chamber and the resulting foam pumped through the hose reel located at the rear of the appliance. Heavy suction pipes were also carried so that the contents of the water tank could be replenished if required, or the whole connected to a mains supply where available.

54B: Basically a motor pump appliance, this superb 60-bhp Austro-Daimler tandem-drive (6 × 4) model of 1936 also carried a foam pre-mixer and other foam generating equipment. Two 176-gpm pumps were carried, the front-mounted unit being a 4-stage model and the other a 30-hp portable unit.

54B

FOAM TENDERS

A popular appliance on the continent of Europe during the Thirties was the 'Universal' type, a multi-purpose machine, often with one fitted pump (usually front-mounted), a portable pump of between 20 and 30 hp and foam pre-mixing apparatus. Thus, when required, the appliance could be operated as an ordinary motor pump.

Later designs were often based on actual fire appliance chassis, using coachwork very similar, if not identical, to the standard motor pump appliance. Where risk was not quite so great, however, foam compound was, and still is, transported in containers on an otherwise standard appliance such as a pump escape or pump water tender.

55A

55C

55B

55A: In 1938 an Austro-FIAT 'AF-Super' multi-purpose machine was the first totally enclosed appliance to serve with the Vienna Fire Brigade. This, like the Austro-Daimler (Fig. 54B), also featured a foam pre-mixer, one portable and one front-mounted pump. The latter was a 6-stage unit of 266-gpm capacity and the portable model a 176-gpm 30-hp TS8 model.

55B: The German Henschel concern built a number of airfield foam tenders in preparation for World War II. One of these was the tandem-drive D1 chassis, equipped with a 550-gallon water tank, 110-gallon foam compound tank and 550-gpm front-mounted pump. After the war many of these, such as the one shown here, were rebuilt for peacetime operation.

55C: This Dennis was also built for operation under wartime conditions. Foam compound was carried in drums (the usual method for British appliances of this type) and introduced into the water supply as shown. The appliance could serve two hoses at any one time. Equipment was by Foamite Limited.

56A

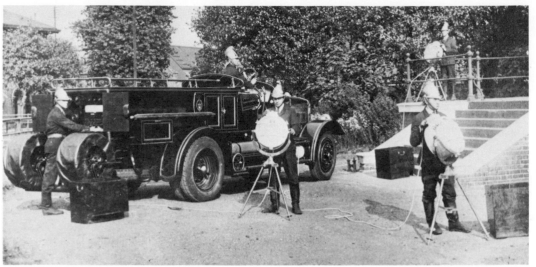

56D

56B

56C

56A: In December 1907 this Knox emergency tender was purchased by the Boston (Massachusetts) Protective Department and installed in the station house of Company No. 3 at 159–161 Roxbury Street. The appliance proved so efficent that in the following year a second machine, identical to the first, was also ordered, and installed in July of that year.

56B: Typical of British emergency tenders of the Twenties was this 19-bhp Dennis supplied to the Brighton Brigade in 1928. Featuring a dynamo and switchboard, its major duty was to transport emergency lighting equipment to major alarm calls.

56C: The London Fire Brigade took delivery of a larger Dennis the following year. Not only did this carry emergency lighting equipment but also a host of other life-saving and rescue aids, including 'Proto' breathing apparatus, 'Novit' oxygen revival set, oxy-acetylene cutting equipment, hydraulic jacks and first-aid box.

56D: This Albion/Merryweather appliance, with 'Braidwood' body, was supplied to São Paulo, Brazil, in 1929. It was specially designed for only two uses—as a transporter and provider of power for emergency lighting equipment, and as an emergency cutting unit using oxy-acetylene apparatus. The latter was carried on the running boards along each side of the vehicle, whilst the former was contained in lockers at the sides and rear of the body. Electric cable was stored on twin reels at the rear.

EMERGENCY TENDERS, DISASTER AND RESCUE TRUCKS

The emergency tender or disaster truck has two basic applications—for use in special cases where fire is not involved (e.g. building collapse, earthquake, etc.) or at major fire outbreaks where specialist equipment is called for. Such equipment includes breathing apparatus, metal-cutting gear, portable pumps, search and floodlights and, where a separate salvage tender is deemed to be unnecessary, salvage equipment also.

The self-propelled emergency tender was introduced soon after World War I. At first it was no more than a commercial van, fitted with a bell or similar warning device, and with special shelves and lockers for stowage of equipment. With the introduction of increasingly complex manufacturing processes, however, and the ever-present threat from new highly

57A

57B

57C

57A: A tandem-drive Steyr Model 440 chassis with 45-bhp engine was the basis of this unusual water-rescue vehicle purchased by the Vienna Fire Authorities in 1935. The boat was powered by a 10-hp Johnson outboard motor and rescue aids included ropes, lifebelts and oxygen revival apparatus.

57B: Meanwhile, a particularly interesting rescue tender had been in operation with the West Ham (London) Brigade for some two years. This had been supplied by Dennis Bros and consisted of a coach-type body mounted on an 80-bhp fire appliance chassis. In addition to its 7-man crew and usual emergency equipment, the new tender carried stretchers, canteen equipment and a 110-volt generator.

57C: The Sunderland Fire Brigade took delivery of a 'one-off' emergency appliance in 1938 based on a 2-ton Bedford chassis converted to forward-control and with coachwork by local coachbuilder, R. H. White & Sons. This joined a similar appliance supplied some four years earlier and amongst the equipment carried were breathing apparatus, extension ladders and a 110-volt generator. Much of this was supplied by John Morris & Sons.

58A: In the same year that Sunderland took delivery of its Bedford, the London Fire Brigade received a new Merryweather machine based, as usual, on an Albion chassis. This was of the low-loading type, carrying such specialized equipment as a smoke extraction unit, the pipes for which can be seen on the roof.

59A: Continental emergency tenders were often even better equipped than their British counterparts. This 9.4-litre Henschel, for example, was fully equipped by Metz and even incorporated a collapsible heavy lifting crane at the rear, of particular use when attending road or rail accidents.

59B: Amongst the many Allied military vehicles which found their way into the hands of Austrian Fire Authorities after the last war was this handsome 4 × 4 Morris-Commercial C8. Such a vehicle, with its all-wheel drive characteristics and ample crew and equipment accommodation, was ideal for use as a rescue tender.

59C: A motor pumper built by the Stutz Fire Engine Company in 1923 has been converted by the Indianapolis Fire Department for use as an emergency tender. Pumping equipment is still intact and major emergency apparatus now carried includes smoke extraction gear, generating equipment and floodlighting.

59A

inflammable materials such as celluloid, more sophisticated tenders were required, heavier than earlier types, and with even more equipment, including generating and switchgear for emergency lighting requirements. Passenger chassis were ideal as their low floor height facilitated easier removal of the heavier equipment than was possible with standard truck chassis.

The rescue truck was a special, and more uncommon, variation of the emergency tender. Such machines were invariably built to the individual requirements of a particular brigade.

59B

Salvage tenders made their appearance back in the days of horse-draught appliances, early designs featuring the familiar 'Braidwood' body. These machines did not carry pumping equipment although their design followed closely that of the normal fire appliance right up to the end of the 1930's. It was then that the now familiar box-bodied salvage tenders came into prominence.

Much of the storage space on such appliances was taken up by jump sheets, tarpaulins, brooms, shovels, portable pumps, etc., although the latter did not become widely accepted until after World War II.

Certain large cities have, for a number of years, been served by a specialist salvage corps, maintained by the insurance companies, whose principle aim is to salvage property from insured premises endangered by fire or other damage. In such cases there is no connection between these and the fire service, other than very close liaison between the two.

61B

61A

61C

60A: In 1911 the Fire Salvage Association of Liverpool took delivery of a fleet of new Dennis salvage tenders. These replaced the Corps' horse-drawn appliances and, like their predecessors, featured 'Braidwood' bodies.

61A: The Underwriters' Fire Patrol of San Francisco was organized in 1875, but did not become affiliated to the City Fire Department until 1943. Amongst the vehicles operated during this time were a couple of American LaFrance salvage tenders, c. 1920 vintage, of which one is seen here outside the Company's premises. Design followed closely that

of the conventional American crew and hose car.

61B: When Liverpool's Dennis tenders reached their retirement, they were replaced by a number of new machines incorporating the bodies taken from the old Dennis appliances. One of these replacements was a Chevrolet, delivered in 1929, and equipped with a new Dennis portable pump and ladder equipment by John Morris.

61C: These, in turn, were replaced after the war by utility Austin auxiliary towing vehicles purchased by the Corps in 1945, which required few modifications before re-entering service.

HOSE-LAYERS

The hose-layer was another specialist appliance developed as a result of rapidly changing conditions. There were two basic types, both of which appeared during the Thirties.

The first transported the hose 'flaked' over special bollards or hangers inside the body. Each hose was coupled to the next and could be paid out through the rear of the appliance at anything up to 30 mph. The other type carried hose 'flaked' in layers along the whole length of the body. Again, the hose was actually laid through the rear of the body while the appliance moved along the road. Both methods remain in use and, apart from one or two slight variations, are more or less international.

Certain 'first-aid' or motor pump appliances have also included hose-laying equipment amongst their apparatus but, again, these are exceptions.

62A

62A: Even before the 1930's there had been attempts to design and build a successful hose-laying vehicle. This unique machine was delivered to São Paulo, Brazil, in 1911 and featured a huge drum onto which the hose was wound. The drum was rotated manually when laying the hose but for rewinding it was chain-driven from the rear countershaft. Although this particular appliance was reasonably successful, the hose-laying method was not. Hose life span was cut to a minimum owing to the fact that wet hose was recoiled onto the drum and not given sufficient time to dry out.

62B: The London Fire Brigade received its first high-speed hose-layer in 1936. Based on a Dennis 'Lancet' passenger chassis, it was specially designed by brigade engineers and could lay one or two lines of hose simultaneously at a speed of approximately 15 mph. Total length carried was 1½ miles.

62B

63A: In 1938 the town of Gosforth, near Newcastle-upon-Tyne, added a dual-purpose machine to its fleet. Using a Bedford chassis, the new appliance combined the capabilities of a motor pump unit with those of a high-speed hose-layer, carrying several hundred feet of hose in a special rear compartment.

63B: British version of the American Ford 91T model was adapted for various fire-fighting applications. This hose-layer carried over a mile of hose, 'flaked' for convenience in four separate compartments. The body underbox contained nozzles, branchpipes and other small gear.

63C: Looking more like an ordinary supplies van than a hose-layer, this Fordson was one of a number of utility units later converted for peacetime operation. Again, hose was 'flaked' along the full length of the body and extra equipment carried in underboxes.

63B

63A

63C

In Europe the late Thirties was a period of re-armament, not only on a fighting front but also in the fire-fighting field. The air-raid precaution, or ARP, vehicle appeared rapidly, often utilizing the capabilities of existing municipal machines such as street-sweepers, gully-emptiers and even refuse collectors. Fire appliances had always been hand-built, indeed they still are, but this took time and that was at a premium. Thus the utility appliance, based upon a standard truck chassis and employing the minimum of materials, was evolved.

As well as utility versions of existing types, two or three new designs were used to combat the new threat. The most famous of these was the British auxiliary towing vehicle, more commonly referred to as the ATV (Fig. 61C). The more 'go-ahead' authorities took delivery of their first ATV's even before the outbreak of war. Based on a 2-ton truck chassis, they resembled ordinary boxvans but included longitudinal crew seating,

65A

64A: With the widespread use of gas during World War I, gas warfare against the civilian population of the British Isles could not be ruled out in World War II. Early ARP appliances, therefore, could be adapted for use as decontamination units as well as emergency fire-fighting appliances. The Scammell street-sprinkler, for example, was fitted with overhead sprays, interchangeable with standard jets, for the dispersal of gas, and with these in action the 1250-gallon tank provided a dispersant solution covering an area 70 ft wide and 35 ft high.

65A: Typical of pre-war ATV's was this 1938 2-ton Morris-Commercial. Pre-war types were generally larger, of more pleasing lines, and incorporated luxuries (such as chromium-plating) not found in later deliveries. The trailer pump was by Dennis Bros.

65B: Another well-known British manufacturer of municipal equipment is Eagle Engineering Limited. They also designed and built a range of ARP machines, including a multi-purpose fire-fighting and decontamination unit of 750 gallons capacity which could be mounted permanently or used as a demountable fixture on any medium size truck chassis. The self-contained 250-gpm pump was driven from an 8-hp Ford auxiliary engine located beneath the operator's seat.

65C: The unique 'Derby' 3-wheeled 'first-aid' appliance, built to Home Office requirements by County Commercial Cars Limited, was frequently nicknamed 'the flea', presumably as a result of its unusual appearance and 'FLE' registration series! Various examples were built, between 1936 and 1938, utilizing an 8- or 10-hp Ford engine for traction and an Austin Seven engine for pumping. Because of its high degree of manoeuvrability, the 'Derby' was frequently assigned to the more densely developed areas of the country. This example was based in Croydon and even included a 'fire dog' amongst its crew.

65B

65C

66A

66B

66C

66D

66A: The Austin 'Twelve' at one time made up 90% of the London taxi fleet. On the morning of 2 September 1939, the authorities commandeered many of these for use as Auxiliary Fire Service towing vehicles. At first they were operated in original form with the addition of ladders, hoses, etc. Later they were repainted in National Fire Service livery and further adapted for brigade use.

66B: During the London 'Blitz' the fire service needed every motor vehicle it could get and occasionally some ingenious ideas were put into practice. Staff of the Finchley AFS Station, for instance, built their own appliance in their spare time. It was based on a secondhand Austin car chassis and was claimed to have cost as little as fifty shillings!

66C: The Austin utility model had many applications. This version carried a Leyland extra heavy pump, again a self-contained unit, and featured side lockers and covered crew accommodation.

66D: The Opel 'Blitz' 2,5—32 shown here was one of 300 similar machines supplied to the German authorities in 1939 for use as ARP motor pumps. Designated type KzS8, pump output was approximately 1760 gpm.

internal lockers, hoses and allied equipment, and some even carried light portable pumps. There was, of course, provision for towing a trailer pump. Other authorities preferred to rely upon the use of high-powered private cars which could be requisitioned when war broke out and some even hired taxi-cabs! However, increasing hire charges and constantly rising repair bills forced them to replace with ATV's by 1941.

In accordance with Home Office specifications, a range of self-contained pump packs, with coupled engine and pump unit, were assembled by selected British manufacturers and allocated as required, some as stationary units in 'high risk' areas, others on trailers, and many more on self-propelled chassis.

Initial batches of British utilities included heavy and extra heavy pump units, of approximately 700 and 1100 gpm output respectively, mobile dam units and pumpless escape carriers. The latter could tow a trailer pump,

67B

67A

67C

67A: Another ingenious 'one-off' appliance was a street fire party trailer assembled by Woolwich commercial traveller Harry Osborne. Costing a mere thirty shillings, the specification included wood salvaged from . . bombed premises, a 40-gallon household water tank and even a couple of dustbin lids for use against incendiaries!

67B: Versatility was the main requirement of the self-contained pump pack. The Dennis heavy pump unit could be mounted on a trailer, such as this Taskers outfit, or . . .

67C: . . . on a truck chassis like the Dennis 'Ace'. Equipment was the same in both cases, with the addition of an extension ladder on the self-propelled unit.

but this arrangement was most unsuccessful, not only because of manoeuvring difficulties, but also because the pump had to be removed before the escape could be unshipped. Later deliveries, therefore, incorporated a centrifugal pump mounted ahead of the road engine and driven off the crankshaft. A small percentage of escape carriers were also equipped with twin hose reels and a 130-gallon water tank, the hose reel pump being driven from a power take-off.

The mobile dam unit was the forerunner of the water tender and trailer pump combination. This, in turn, was later developed into the post-war pump water tender. There were many variations. Early types were converted from pre-war civilian vehicles but later deliveries were based upon utility truck chassis with dropside bodies. Each unit carried a demountable canvas or easily assembled steel tank of up to 1000 gallons capacity, various lengths of hose, and usually towed a trailer pump.

68A

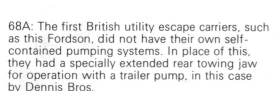

68A: The first British utility escape carriers, such as this Fordson, did not have their own self-contained pumping systems. In place of this, they had a specially extended rear towing jaw for operation with a trailer pump, in this case by Dennis Bros.

68B: The earliest mobile dam units were often conversions of earlier civilian vehicles. This converted coach was typical. The entire rear half of the body was cut away to house the demountable steel-framed canvas dam and Scammell 475–500-gpm trailer pump.

68B

AUXILIARY AND UTILITY TYPES

69A: Within a couple of years the mobile dam unit consisted, generally, of a commercial truck which could undertake work vital to the nation's economy during the day and act as a fire support vehicle at night. The Dodge Model 82 3/4 tonner was one of the most popular of this type and often carried the larger dams of up to 1000 gallons capacity.

69B: Built in 1943, the Opel 'Blitz' 3,6–36S combined water tender and pump unit was the German equivalent of the British water tender. The tank was of 550 gallons capacity and pump output 330 gpm.

69A

Water tenders were delivered a little later. The bulk of these were fitted with a 500–600-gallon tank, one or two hose reels and, usually, a PTO-driven pump, often supplemented by a trailer pump. Such appliances were the mainstay of the German Fire Service at this time. An interesting variation on this theme was the combination flat truck/water tender supplied in small numbers at the very beginning of the war. The pioneer of this design was the Sparshatt Company which developed a 400–gallon model with the tank built into the body below floor height. Thus, the vehicle could be used for carrying goods and general supplies and, at the same time, be available for use in case of enemy attack.

Utility turntable ladders, hand- or mechanically operated, replaced standard pre-war types damaged by enemy action. Some had front-mounted pumps, while others had none at all. The wartime version of the hose-layer was the hose lorry, again a normal commercial model, but

69B

generally with specially designed box body for paying out hose from the rear. Because British manufacturers had difficulty in supplying sufficient hose for the war effort, these, and many other appliances, operated at reduced capacity owing to the use of imported heavy, cumbrous, cotton-jacketed rubber hose.

The coming of peace saw huge fire service replacement programmes. Even so, large numbers of utility machines were rebuilt for 'civilian' operation and many German appliances of this type are still in regular operation on the continent of Europe. There can be no doubt that the period of World War II had a very marked effect indeed upon the development of fire-fighting equipment.

70A

70B

70C

70A: The Italian Alfa-Romeo Model 800, built between 1940 and 1943, saw service in all fields during hostilities. As a utility turntable ladder, however, it was not so common. Note the crew compartment in the rear of the cab.

70B: The Hansa-Lloyd marque was manufactured by Carl F. W. Borgward GmbH until 1945 when all models became known as Borgward. A few Hansa-Lloyds were supplied to the German Fire Service in 1939. These were equipped with hand-operated Magirus 60-ft ladders and after the war passed to Occupation Forces, as shown.

70C: The most common British utility TL was based on the Austin chassis. Because German equipment was no longer available British manufacturers, notably Merryweather & Sons, turned their attention to this type of appliance, building them in increasing numbers as the war progressed. This 60-ft mechanically operated Merryweather had its own front-mounted pump and could double as a water-tower if required.

71A

71B

71A: Examples of the ingenuity of individual brigades were rife throughout the war years. The smaller brigades, and those situated where 'fire risk' was not great, often relied upon improvisation rather than specialist machines. One such example was this unique trailer used by the Gosforth ARP Unit. Hose was stowed in three compartments, located one above the other, and paid out at the rear. The towing unit was an 8-cylinder Buick.

71B: Very useful conversion work was undertaken on a number of high-powered private cars requisitioned for fire fighting duty early in the war. This 1933 Terraplane saloon was used as a foam tender. It was equipped with a 120-gpm built-in pump, foam nozzles, ladders and trailer towing equipment.

71C: Utility versions of the emergency tender, like this austere example based on a Dennis 'Falcon' passenger chassis, found ample work on bomb damage salvage and rescue duties, particularly during rocket attacks towards the end of the war.

71D: Even after the war a number of countries continued to maintain an Auxiliary Fire Service. The United Kingdom was no exception, maintaining this until recent years. An unusual light 'first-aid' appliance such as this would carry numerous small tools and items of equipment, notably for use in rescue work. Based on a 10/12-cwt Bedford, the window conversion was undertaken by Messrs Martin Walter of Folkestone, Kent.

71D

71C

INDUSTRIAL APPLIANCES

72A

72B

72C

72D

72A: Amongst the more unorthodox industrial types built by Dennis Bros was a hand-drawn pumping unit supplied to the Great Eastern Railway in 1912. The 300-gpm Gwynne pump was powered by a 45-hp 4-cylinder engine.

72B: In 1936 R. A. Lister & Company Limited adapted their popular 'Auto-Truck' works truck type RHC/NU9G for use as a 'first-aid' fire appliance. This was offered for sale as an ideal air-raid precaution appliance, consisting of the JAP-engined truck, a light 2-wheeled handcart, and a set of hand extinguishers and hoses, the latter for coupling direct to the mains supply.

72C: The Lister Company also operated its own works brigade. One of the appliances was a smart British-built Ford Model 51 of 1936. The other machine, a 1926 Morris-Commercial seen in the background, was sold to the Historic Commercial Vehicle Club in 1970.

72D: The Durham & Northumberland Collieries Fire & Rescue Brigade specified 6-wheeled Leyland 'Terrier' van-type appliances when ordering four new machines in 1936. Equipment included a Gwynne 2-stage pump of 700–900 gpm output located amidships, removable searchlight with cable and a multitude of other rescue aids.

Industrial premises are amongst the most vulnerable, and frequently the most dangerous, when fire occurs. It is not surprising, therefore, that the majority of large industrial concerns, and some of the smaller ones, maintain their own private brigades.

The first British brigades of this type sprang up during the 1840's, a period when fire prevention in industry had difficulty in keeping pace with new manufacturing processes and resultant products. Most famous of these brigades were those at Hodges' Distillery and Brown, Lennox & Company, both London-based and both of which also turned out to fires other than on their own premises.

Lightweight appliances have always been especially popular for this class of fire-fighting. Industrial areas have never been renowned for ease of access, so fire-fighting equipment has had to be tailor-made for the area in which it is expected to work. Thus, the handcart, the light works truck and special 3-wheeled and car-based machines have been particular favourites in this field.

Collieries and similar concerns have also operated their own fire and rescue services, employing special vehicles for the carriage of breathing apparatus, gas dispersant squads, etc., and capable of traversing rough country at high speeds. All have found the industrial appliance indispensable both for saving lives and for salvaging machines and property.

73A

73B

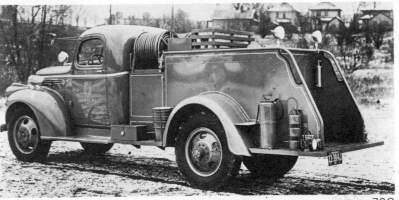

73C

73A: Even the famous Scammell mechanical horse was adapted for use as an industrial and ARP appliance. Because of its excellent manoeuvrability, such a machine was ideal for industrial fire-fighting. This version featured a 350-gallon water tank, twin hose reels and 50-gpm PTO-driven pump.
73B: A particularly rare fire-fighter was this ex-German Army Stoewer 4 x 4 model operated by Vogel & Noot, of Wartberg, Murz, Austria. This was used mainly as a hose-carrier rather than an actual fire appliance.
73C: The John Bean Manufacturing Company, a Division of the Food Machinery Corporation, adapted this Chevrolet 1½-tonner for their own use in 1942. Although designed primarily for use as an industrial chemical 'fog' fire-fighter, manned by John Bean's factory personnel, it was even used outside Company premises from time to time.

AIRFIELD FIRE CRASH TENDERS

As the use of civil and military aircraft increased during the Twenties and Thirties so, too, did the need for rapid extinction of aircraft fires, many of which could be as dangerous as a serious oil fire, with the added threat of possible loss of life.

At first more or less standard appliances, with 'Braidwood' bodywork and frequently inferior or inadequate equipment, were adapted for airfield operation, but it became increasingly obvious, as the amount of fuel and numbers of passengers carried increased, that there was a very real need

74B

74A

74C

74A: Between 1934 and 1938 the German authorities were preparing for hostilities. New designs and improved fire-fighting methods were developed, not least of which was the carbon dioxide gas airfield fire crash tender. This compact design, based on a tandem-drive Mercedes-Benz LG3000 chassis, incorporated numerous hose reels, each linked to its own CO_2 supply.

74B: Unipower Limited (formerly Universal Power Drives Limited), of Perivale, Middlesex, are well known for their third axle conversion work. One of the rarer examples was this Fordson for the Royal Air Force, with open cab and carbon dioxide extinguishing apparatus. It will be noted that each set of gas cylinders, as in the Mercedes, had its own delivery hose.

74C: It was not surprising that when the Municipal Airport was opened at Luton, Bedfordshire, in 1938, a locally built Bedford 30-cwt model was supplied for use as crash tender. It was of most unusual design, being supplied in chassis-scuttle form with flat body and Foamite foam generating apparatus.

74D: Typical of so many early British fire crash tenders was the Commer/Merryweather LN5 built in 1939, of which a number were supplied both at home and abroad. Equipped with extension ladder, foam and carbon dioxide equipment, it was capable of laying a 100 sq. yd 'carpet' of foam $2\frac{1}{2}$ in thick in 6—8 minutes.

74D

for a highly specialized appliance developed specifically for airfield work.

Germany was a leader in this field, developing the first totally enclosed airfield fire crash tenders about 1935. In the United Kingdom, right up to the close of World War II in fact, the open-bodied and particularly the 'Braidwood' style was believed to be the more efficient, however.

During the war the American Cardox Corporation, now part of the Chemetron Corporation, developed an entirely new type of airfield fire crash tender, which was to revolutionize the fighting of aircraft fires. The basis of this was a large tank containing some 3 tons of carbon dioxide and a special foam compound. These were ejected, individually or in combination, onto the blazing wreckage by means of up to six nozzles, including both an overhead boom and a ground level spray. A similar principle, but employing foam 'monitors', is used to this day.

By 1950 the all-wheel drive chassis had come into its own as the standard for fire crash tender applications. It was ideal for cross-country operation and since then still more all-wheel drive units have been developed specifically for this type of work.

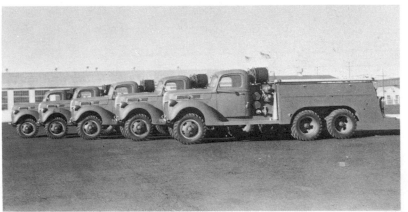

75B

75A

75A: The Crossley 'Q'-Type 4 × 4 appliance was built for wartime operation with the Royal Air Force. This, the long-wheelbase version, was equipped with a 1000-gpm pump, 300-gallon water tank and 25-gallon foam compound tank.

75B: A fleet of Marmon-Herrington 2½-ton 6 × 6 conversions of the American Ford, Model MM5–6, was supplied to the Dutch Government for use in Java in 1941. Bodies were of the open type, incorporating both CO_2 and foam equipment.

75C: The Mercedes-Benz L4500A 4½-ton 4 × 4, powered by a 112-bhp diesel engine, appeared in 1941. One application was as a foam tender with canvas cab for airfield operation. With such a low silhouette and a removable canvas cab roof, it would appear that this vehicle could also be transported by air if necessary.

75C

76A: The special carbon dioxide and foam equipment developed by the American Cardox Corporation was mounted on a 6 × 6 Sterling DDS 235 or Reo 29-FF chassis. All operations were controlled from the cab, this being the Reo version. Note the ground-level spray at the front of the vehicle.

77A: For the large scale generation of foam the Air Ministry developed an experimental crash tender based on an AEC (O)854 6 × 6 chassis. Fire-fighting equipment was supplied by Pyrene and consisted of twin tanks, each carrying its own foam 'monitor', and compressed air and carbon dioxide apparatus for subsidiary work at the rear.

77B: The first crash tender designed and built by Pyrene themselves was based on an ex-military Bedford QL Model all-wheel drive chassis and supplied to the States of Jersey in 1947. The specification called for a 500-gallon water tank, container for 30 gallons of foam compound, and CO_2 extinguishing apparatus. The pump was a 75-hp Coventry Climax 380–400 gpm unit.

77C: The first 4 × 4 Thornycroft 'Nubian' crash tenders appeared some two years later. Pyrene built a number of these machines up to 1962, based mainly on similar specifications. Fire-fighting media consisted of 500 gallons of water, 60 gallons of foam compound, and CO_2 equipment. Water was delivered at the rate of 380 gpm and a novel feature was the stowage of hose and nozzles in boxes on either side of the radiator.

77A

77B

77C

FOREST FIRE APPLIANCES

Where there are large areas of forest, bush or scrubland there will always be an acute fire risk, particularly where there is a low rainfall and high temperatures. California, the South of France and the Australian 'bush' are cases in point.

For a number of years scientists have been perfecting ways of conquering the fires in these areas, and have now resorted to the use of aircraft dropping 'fire bombs' (large amounts of fire-choking chemicals) in the heart of the flames. As long ago as the 1920's it was realized that the raging infernos prevalent in these areas, driven on by terrific winds generated by the searing heat, could not be controlled by the usual fire-fighting vehicles and methods. Normal appliances could get only as close as the nearest road or track, and all too frequently these were obliterated by the flames.

Top priority was given to the fact that fire-fighting equipment for use in such cases had to be capable of traversing the roughest of terrains. Thus, during the early Thirties the half-track and, later, the full-track type appliance was utilized. After World War II the glut of ex-military vehicles provided a ready source of all-wheel drive models and the tracked type was largely forgotten for this type of work.

79A

79C

79B

78A: One of the earliest custom-built appliances for forest fire work was a prototype in 1933 by Thirion of Le Havre, based on a Citroën-Kégresse half-track. This appliance was designed to spray water over a large area, but it delivered too much water in too short a time.

79A: This 36-bhp Austro-Daimler of 1929 was specially equipped to deal with forest fires, carrying fire hooks, brooms, shovels, pick-axes and even stirrup pumps.

79B: A 1944 ex-US Army GMC Model CCKW-353 6 × 6 was the basis of this machine still operating in Soest in the Netherlands. Its main duty was to extinguish heath fires but it was also maintained as a 'second line' appliance. Converted in 1949, its equipment included three hose reels, a 656-gallon water tank and Bean high-pressure pump.

79C: This German trailer appliance of World War II carried soda-acid extinguishers, a 2-man saw, and special 'beaters' for extinguishing perimeter outbreaks.

79D: Since 1947 a number of French manufacturers have based heavy forest fire appliances on ex-Services 6 × 6 truck chassis. This GMC Type CCKW-353, for example, was adapted by Guinard. It included a 770-gallon water tank, rear hose reel and portable pump amidships.

79D

INDEX

ACKNOWLEDGEMENTS

Credit is hereby given to all manufacturers, brigades and other bodies who have assisted in the compilation of this book. In particular we should like to thank Merryweather & Sons Limited, Dennis Bros Limited, the Trustees of The London Museum, The London Science Museum, The London Fire Brigade, The Pyrene Company Limited, the Chief Illustrator (Home Office Illustration Branch), Mr Arthur Ingram, The Commercial Vehicle and Road Transport Club, the B. H. Vanderveen Collection, Mr Larry Auten, Mr Alfred Krenn, Mr Frederick Lane and Mr J. G. Jeudy of 'Protection Civile et Securité Industrielle'.